Discover the

BETA GLUCAN

Secret!

for Immune Enhancement
Cancer Prevention & Treatment
Cholesterol Reduction
Glucose Regulation

& much more!

A Health Learning Handbook

Beth M. Ley, Ph.D.

BL Publications
Detroit Lakes, MN

BL Publications, Detroit Lakes, MN
For orders call 1-877-BOOKS11
email: blpub@tekstar.com

Library of Congress Cataloging-in-Publication Data

Ley, Beth M., 1964-
  Discover the beta glucan secret! : for immune enhancement, cancer prevention & treatment, cholesterol reduction, glucose regulation, and much more! : a health learning handbook / Beth M. Ley.-- Rev. and updated, 2nd ed.
        p. cm.
Includes bibliographical references and index.
  ISBN 1-890766-18-6
  1. Glucans--Health aspects. 2. Glucans--Therapeutic use. I. Title.
  QP702.G48 L49 2001
  615'.329562--dc21

                                          00-013128

Printed in the United States of America

This book is not intended as medical advice. Its purpose is solely educational. Please consult your healthcare professional for all health problems.

Credits

Cover Design: BL Publications/Jerrita Gronewold

Proofreading: Deborah Brenk

---

*Use your food as your medicine
and your medicine as your food.*

Hippocrates

Table of Contents

YOU NEED TO KNOW...
THE HEALTH MESSAGE

Do you not know that you are God's temple and that God's Spirit dwells in you? If anyone destroys God's temple, God will destroy him, For God's temple is holy and that temple you are.
1 Corinthians 3:16-17

So, whether you eat or drink, or whatever you do, do all to the glory of God.
1 Corinthians 10:31

Introduction

Beta glucan is found in all living organisms. It is a unique natural compound derived from the cell walls of yeast. Beta glucan triggers an immune response in the body creating a system of defense against viral, bacterial, fungal, parasitic or potentially cancerous invaders. It also has potent anti-oxidant and free radical scavenging capabilities.

The Immune System

The basic function of the immune system is to protect us against infection, illness and disease of all kinds. It fights off thousands of predatory environmental and infectious microorganisms which can invade and damage virtually every part of the body. The immune system expels pathogens, toxic chemicals and tumorous cells generated through mutation. The immune system also aids the body in tissue repair and healing. It's goal is to maintain homeostasis (balance) in the body.

The immune system involves a finely-tuned, highly-integrated cascade of events that destroy a would-be invader. Once the immune system identifies an unwanted guest, it brings an awesome array of chemical and cellular weapons to act towards its elimination. This non-specific, intricate and elaborate immune response with built-in checks and balances, can readily discriminate between the "self" and "non-self," that is, the invading pathogens. The intricate network of the immune response ensures that the body does not turn on itself and lose the ability to distinguish between self and nonself.

Autoimmunity

Loss of the ability to distinguish between self and nonself can lead to autoimmunity, and the body may manufacture antibodies and T-cells directed against its own cells, cell components or specific organs. Autoimmune conditions include allergies, asthma, rheumatoid arthritis, dermatitis, diabetes, Systemic Lupus Erythematosus (Lupus), Multiple Sclerosis (MS), fibromyalgia and also may underlie numerous additional chronic diseases. Overstimulation caused by constant exposure to various pathogens, may also result in autoimmunity.

Weakened Immunity

When the immune system is working properly, we remain healthy. However, the immune system can and does become compromised. This may be a result of constant environmental assault (such as from exposure to pseudo-estrogens, pesticides and pollutants in air, food and water, certain medications and various other toxins), persistent metabolic damage (for example, high free radical stress), poor nutrition, chronic infection or advancing age. Strenuous physical activity and emotional stress are also known to adversely affect immunity. Numerous diverse factors may compromise the immune system and disturb its various components exquisitely balanced.

It is no secret that our immune systems are under huge amounts of stress today. This excess stress weakens the immune system making us more vulnerable to illness. When our bodies become overwhelmed, the regulatory features of the immune system weakens and becomes less effective resulting in:

- Fatigue, loss of stamina and energy
- Frequent colds and infections
- Loss of appetite and weight loss
- Fever and night sweats
- Skin rashes and cold sores
- Diarrhea
- Increased severity in allergy symptoms
- Swollen lymph nodes, etc.

In time, this can result in a more serious immuno-compromised condition (such as chronic fatigue, cancer, HIV, etc.). It is widely believed that cancer is caused by a decline in the immune system caused by aging and other factors. The AIDS virus destroys immuno cells and triggers diseases.

Over time, cell-mediated immunity may become inadequate or malfunction. This could allow multiple genetic mutations in the same location (malignant transformations) to reach cell proportions resulting in abnormal or malignant growth. The process is generally quite slow and it may take many years and many bouts of illness to truly manifest itself.

The therapeutic immune-stimulating benefits of beta glucan include a wide range of health problems ranging from cancer, chronic fatigue syndrome, high blood pressure, rheumatoid arthritis, diabetes, hepatitis, high cholesterol, tumors, HIV and AIDS.

To maintain good health, minimize the frequency and severity of all illnesses, and recover quickly and enjoy a healthy quality of life, it is imperative that the immune system is functioning efficiently and optimally. The more detriment we expose the body to, the harder the immune system has to work to maintain our health.

For all these reasons, natural immune system support is very important, making beta glucan an excellent supplement for all of us.

Polysaccharides and Beta Glucan

Polysaccharides (high-molecular-weight complex sugars), most notably beta glucan, have attracted research attention for their immune-enhancing and/or tumor-retarding affects in the body, and their ability to help regulate cholesterol and glucose levels in the body. Researchers consider Beta-1.3 and 1.6-glucan some of the most powerful polysaccharides.

Beta glucan is found in every living organism. Beta glucan is a non-digestible long-chain carbohydrate (which could also be referred to as a soluble fiber) found in the cell walls of certain plants.

Glucans are characterized by repeating glucose subunits jointed by a beta-linkage between various carbons of the glucose ring. The linkages primarily proven beneficial are the beta-1.3 and 1.6 linkages, although in nature they are found bound together as a sugar/protein complex.

Beta Mannen is a mannoprotein (Acemannen, i.e. Aloe vera) and contains beta-1.6 glucan. It has hormonal effects and similar growth factors that are selective to the Langerhans (immune cells). It directly induces the structural integrity of cells, alertness and a number of circulating immune cells. Glucan, as a sugar molecule, stimulates and energizes immune cell cascades such as macrophages, neutrophils, Natural Killer cells, Tumor Necrosis factors, etc. into action while Mannoproteins build structural integrity.

These polysaccharides activate various immune effector cells to attack tumor cells, but also potentiate the activities of various immune mediators, including lymphokines, and interleuken-1 and interleuken-2. Lymphokines and interleuken-1 and -2 are important components of white blood cells that regulate the human immune system. Since beta glucan stimulates these cells, they exert a strong anti-tumor and regulatory effect upon the immune system.

Regulation of these immune cells is important not only to the possible prevention and treatment of cancer and tumors, but for other immune dysfunctions as well. For example, medicinal mushrooms produce anti-viral activities and help regulate the immune system during chemotherapy and other highly stressful events.

In most cases, oral ingestion is more effective than direct injection. The polysaccharide complexes are partially broken down by the action of acids and enzymes in the digestive system, making them more readily available.

Beta glucan may benefit an imbalanced immune system in several different ways. Beta glucan triggers an immune response in the body creating a system of defense against viral, bacterial, fungal, parasitic or potentially cancerous invaders. It also has potent anti-oxidant and free radical scavenging capabilities.

Beta glucan is water soluble and effective when taken orally. There is no known toxicity or side effects from using beta glucan.

The structure of immune-enhancing beta glucan is termed as beta-1,3 or 1,6-glucan. This means that glucose molecules are linked together in a so-called beta-1,3 chain, which forms the backbone of the beta-glucan, with glucose molecules branching out from the backbone in beta-1,6 linkages.

These factors decrease the body's ability to ward off infection and disease. Less active defensive cells and lowered antibody production increase susceptibility to chronic infections by yeast, bacteria and viruses and slowed wound healing, while increasing the risk of cancer and degenerative diseases such as arthritis.

beta glucan

Macrophage phagocytosis (cell eating) "clean up" activity is activated by beta glucan.

Beta-glucan is well known to researchers as a supplement to improve immune function. The most important property of beta-glucan is that it is an important immune activator. It increases nonspecific immunity and fosters resistance to infection. It does this by targeting and activating immune cells, primarily macrophages, but also other white blood cells including monocytes, neutrophils and precursor immune cells.

Macrophages, which means "big eaters," devour invading pathogens of all sorts. The macrophage cell is a marvelous single-cell organism which can identify and engulf, by process of phagocytosis (cell eating), any abnormal cells it may find in the human body. Inside the

> *Beta glucan activates macrophages to identify and eliminate invading pathogens from the body, triggering a chain reaction leading to increased immune activity.*

cell are digestive enzymes called lysozymes that break down the substance rendering it harmless.

Macrophage cells do a great job of eliminating harmful cells from our bodies, but only when they are motivated to do so. Macrophage cell membranes possess specific receptors (or docking sites) for beta glucan. Upon binding of beta glucan to this receptor, the macrophages are activated to identify and eliminate invading pathogens from the body. Essentially, beta glucan triggers a chain reaction leading to increased immune activity. As an example, beta glucan increases phagocytic activity by engulfing foreign or dangerous cells (bacteria, viruses, endotoxins and debris), leading to their destruction and elimination. (Paulik)

Beta glucan benefits specific white blood cells of the immune system, especially macrophages. These white blood cell scavengers form a first line of defense against invaders and act as internal regulators of the immune system.

Macrophages also produce cytokines that activate and mobilize the immune system and they also increase bone marrow production of immune cells. Beta glucan induces expression of cytokines which jump start the immune system in case of a precipitous threat.

Beta glucan also stimulates the production of interferon and interleukins to gear up defenses by activating T cells and other immune system cells. It can trigger the release of tumor necrosis factor (TNF), a potent defensive protein, and superoxide for human blood cells such as monocytes, to wage chemical warfare against invaders. (Hoffman)

Reduced bone marrow proliferation implies lowered white cell counts and an increased risk of infection and cancer. Therefore, it is significant that beta glucan stimulates production of immune cells by the bone marrow.

Other studies offer the prospect of protection against radiation damage: When mice were exposed to increasing levels of radiation, beta glucan administered pre- and post-irradiation decreased mortality. Stimulation of bone marrow cell production was a major benefit. In other studies, human natural killer cells, responsible for destroying cancer cells and other potentially harmful cells, were activated in the presence of beta glucan.

Immune System Enhancing Effects of beta glucan:

- Increases production of cytokines

- Stimulates production of Interferon and other immune system enhancers

- Increases phagocytosis activity

- Activates natural killer cells

- Stimulates the production of pro-inflammatory signal proteins such as interleukin-1

- Activates T cells and other immune system cells

- Triggers release of tumor necrosis factor (TNF)

Fighting Cancer

While cancer is among the top two causes of death today, the most common types of cancer are largely *preventable*. Chemical and environmental factors may be responsible for 90% of all cancers. Exposure to these factors is greatly our own responsibility. Additional cancer risk factors that are more difficult to control include genetics, age, hormone imbalances and toxic buildup in the cells, especially the colon.

Cancer is the abnormal growth of cells in our bodies. Cancer originates as a result of normal cell mutation through its genetic chromosome material, RNA and DNA. Normally, cells replicate themselves continually at a rate synchronous with normal growth and repair in a manner specific for its purpose in the body. A cancerous cell multiplies faster than it should and loses normal differentiation.

The actual mechanisms by which cancer is caused is still speculative. It involves a weakening and breakdown of the immune system. At any given time, there are literally thousands of potentially "precancerous" mutated cells throughout the body. A healthy immune system is able to effectively locate and eliminate them from the

"The broad spectrum of immunopharmacological activities of beta glucan includes not only the modification of certain bacterial, fungal, viral, and parasitic infection, but also inhibition of tumor growth."

Nicholas DiLuzio, Ph.D., Department of Physiology, Tulane University School of Medicine body. If it does not, the mutated cells turn into cancer.

We know cancer prevention is the best way to fight cancer. We are told not to smoke or drink and to avoid foods containing nitrates, pesticides and other chemicals, to avoid stress and even the sun (at least without sun-screen). We know these things weaken our most important defense against cancer — the immune system. Maintaining a healthy intracellular environment helps prevent cancer, and correcting a toxic environment may lead to successful non-toxic treatment for cancer.

Beta glucan has been intensively studied since the 1950s for its anti-tumor and immuno-stimulating properties. The clinical usefulness of Beta-1,3/1-6-Glucan has been demonstrated with individuals with various forms of cancer.

Several research studies show that beta glucan provides anti-tumor activity in mammals. Beta glucan vitalizes production of interferon and interleukin in small animals. This effect indirectly functions to destroy or prevent the proliferation of cancer cells via a cytokine inducing effect.

A number of Oriental mushrooms are well known for their cancer-fighting properties in the body including the Brazilian Agaricus blazei Murril (having the highest concentration of beta glucan), Grifola umbellate, and Phellinus yucatensis.

In 1980, the Japan Cancer Association showed that the beta glucan of Agaricus is effective against various types of cancer cells including: Ehrlich's ascites carcinoma, sigmoid colonic cancer, ovarian cancer, breast cancer, lung cancer and liver cancer as well as against solid tumors.

More recently, Japanese researchers at the

Department of Pharmacology, Mie University School of Medicine examined the anti-tumor activity of beta glucan prepared from Agaricus against four kinds of established mouse tumors. They found it to be effective against all four types of tumors. (Ito)

Researchers have deduced that the beta glucan extract activates the immunity of normal biological tissue, so that even when a virus or other external factors enter the tissue, macrophage and interferon production within the tissue is vitalized to prevent the multiplication, metastasis and reoccurrence of cancer cells. Results of the experiments suggest that it also activates metabolism by revitalizing normal biological tissue. (Osaki)

Researchers at the Department of Pathology, University of Louisville, KY report the potential of immunotherapy for cancer using soluble yeast beta-glucan to override the normal resistance of certain tumor cells to the cytotoxic activation of phagocyte and NK cell CR3, allowing this important effector mechanism of the C system to function against tumor cells in the same way that it normally functions against bacteria and yeast. They also pointed out that beta-glucan can enhance tumor-localized secretion of the cytokines TNF-alpha, IFN-alpha, IFN-gamma and IL-6 to also fight the tumor. (Ross)

Extensive clinical studies of Oriental mushrooms for colorectal and breast cancer patients are now underway.

Beta glucan can be used alone or in combination with chemotherapy and/or radiotherapy. As a result, it was observed that the immunological parameters of the patients improved, and the survival periods have been considerably prolonged.

Cholesterol Reduction

Studies show that a diet high in polysaccharides containing beta glucan reduces the blood level of cholesterol. Beta glucan, as a soluble dietary fiber, forms a viscous gel in the digestive tract aiding the regulation of cholesterol. Oat and barley beta glucan (4-13 grams/daily) have also demonstrated to decrease LDL cholesterol, the less desirable form of cholesterol, in individuals with high blood cholesterol levels. (Uusitupa Braaten, Lia, Bobek, Lovegrove)

NOTE: *Studies demonstrate that beta glucan from oats also possess immunomodulatory activities capable of stimulating immune functions. (Estrada)*

Recent results with beta glucan of mushroom origin (Pleurotus ostreatus) were statistically very promising. There was a 12.6% decrease of very-low-density lipoproteins (VLDL) and of LDL cholesterol in the serum and 27.2% decrease in triglycerols. After the seventh week of the experiment, the cholesterolemia was lowered by almost 40% as compared with control animals kept on the same diet but without the mushroom. (Bobek)

The dried and powdered mushroom (Pleurotus ostreatus) fed to hamsters at 2% of a high-fat diet for six months lowered serum VLDL in blood plasma as well as cholesterol and triglycerol levels in the liver. Serum VLDL decreased by 65-80% and total serum lipid levels were reduced by 40%. Beta glucan from the mushroom as part (4%) of a normal diet was found to lower serum and liver levels of cholesterol in hamsters after two months. As 5% of a high fat diet in rats, P. ostreatus lowered serum cholesterol accumulation by 45% in three months. (Bobek)

Glucose Regulation

When consumed with a meal, beta glucan decreases the uptake of glucose (lowers the glycemic index of the food) and thus supports blood sugar regulation. (Wood, Tappy, Wursch)

Researchers at the Division of Endocrinology and Metabolism, Ottawa Civic Hospital, University of Ottawa, Ontario, Canada wanted to characterize the effects of isolated and natural sources of soluble beta-glucan, when incorporated into a complete meal in diabetics. They found that oat bran and wheat farina plus oat gum meals reduced the post meal plasma glucose excursions and insulin levels when compared with the controls. This study shows that both the native cell wall fibre of oat bran and isolated oat gum, when incorporated into a meal, act similarly by lowering post meal plasma glucose and insulin levels. A diet rich in beta-glucan may therefore be of great benefit in the regulation of post meal plasma glucose levels in subjects with Type 2 Diabetes. (Braaten)

According to researchers at Nestle Research Centre, Lausanne, Switzerland, "*A 50% reduction in glycemic peak can be achieved with a concentration of 10% beta-glucan in a cereal food. A significant lowering of plasma LDL cholesterol concentrations can also be anticipated with the daily consumption of > 3 grams of beta glucan. Diabetic individuals can benefit from diets that are high in beta-glucan, which, as a component of oats and barley, can be incorporated into breakfast cereals and other products.*" (Wursch)

Asthma and Adult Respiratory Distress Syndrome

Interesting data from researchers at the Mayo Clinic (Thoracic Disease Research Unit) suggest that beta glucan may play an immunomodulatory role for individuals with asthma and lung disorders such as adult respiratory distress syndrome.

Beta glucan, depending on the dose, has both TNF alpha-stimulating and -inhibiting effects of fungal beta-glucans during infection. Too low a dose can actually stimulate TNF alpha secretion from macrophages, so the proper dose is an important factor.

Tumor necrosis factor-alpha (TNF alpha) is a potent cytokine believed to participate in the development of endotoxin-induced shock, in the triggering of an asthma attack and of lung disorders such as adult respiratory distress syndrome. (Hoffman)

Because human studies are not yet available on asthma or other lung conditions, it is difficult to recommend a dosage. If after a few days on an initial dosage, you feel that your lung condition has worsened, increase your dose.

NOTE: *Lung distress has not been reported with using beta glucan, this is just a precautionary informational message in accordance with the research data that is available.*

Anti-inflammatory Benefits

Animal studies conducted at the Department of Veterinary Molecular Biology, Montana State University, Bozeman demonstrated the anti-inflammatory regulatory benefits of beta-(1,6)-branched beta-(1,3)-glucan (soluble beta-glucan).

The overall cytokine pattern of leukocytes from soluble beta-glucan-treated mice reflects suppressed production of inflammatory cytokines, especially TNF-alpha. Taken together, our results suggest that treatment with soluble beta-glucan can modulate the induction cytokines during sepsis. (Soltys) Other researchers, including those at the Tokyo University of Pharmacy and Life Science, have also shown the anti-inflammatory potential of beta glucan. (Sakurai)

This is consistent with the suggestion that beta glucan may be helpful for asthma, but also suggests it's potential for numerous other conditions such as allergies, dermatitis, tendonitis, bursitis, arthritis and lupus.

Some clinicians report in their experience that some of their patients with inflammatory auto immune conditions (such as arthritis, lupus and asthma) may experience a temporary inflammatory reaction upon starting beta glucan therapy. One may feel as though their symptoms are increasing instead of improving. This will pass after a few days and then the symptoms will greatly improve. MSM is also recommended in conjunction with beta glucan for individuals with joint disorders such as arthritis or bursitis. (Borman)

Topical Use

Polysaccharides have been used in creams, ointments, suspensions and powders for topical applications all around the world by scientists and brand name cosmetic companies found in major department stores. Beta glucan is usually present in low concentrations ranging from 0.5-2.00%. These topical applications show water-binding, moisturizing and anti-inflammatory effect.

Clinical evaluation of beta glucan preparations have been applied 1-2 times a day for a period of 2-4 weeks with the following results:

- In 15 patients with dermatitis, a cream containing beta glucan was proven effective in 53% cases and very effective in 47%.

- In 15 patients with eczema, a cream containing beta glucan was proven most effective in 60% of the cases and very effective in 34% and ineffective in 6% of the cases.

- In 15 patients with dermatitis, a body lotion containing beta glucan was proven most effective in 60% and very effective in 40% of the cases.

- Other benefits of the beta glucan treatments were improved elasticity and hydration and a decrease of wrinkles in the treated skin.

- No side effects were noted. (Zelenkova)

Maintain Optimal Health with Beta Glucan

Many people can benefit from beta glucan and other immune enhancing supplements and activities such as those with:

- Impaired immune systems
- An auto-immune condition (arthritis, lupus, allergies, asthma, M.S., etc.)
- HIV infection
- Cancer who are undergoing radiation or chemotherapy
- Poor nutrition or who eat a diet high in refined foods and preservatives
- High physical or emotional stress (including work related)
- Cardiovascular disease
- Elevated cholesterol
- Diabetes
- Aging concerns

Individuals such as athletes who participate in strenuous physical activity can also benefit from beta glucan supplementation.

Sources of Beta Glucan

Certain plants and microorganisms are naturally high in beta glucan. It can be isolated from the common bakers' yeast (Saccharomyces cerevisiae), various Oriental mushrooms, algae and other seaweed, grasses, oats, barley and aloe vera.

The richest concentrated source is believed to be baker's yeast cell walls. Extraction methods include sodium bath (change the pH and explode the cell from within), crushing with glass beads, solvent extraction with alcohols and benzenes, etc. When the "insides" of the cells are removed, only the walls are left. It typically may require 40 pounds of walls to get one pound of glucan extract concentrate. It is finely ground and filtered, leaving the valuable fatty acid matrix intact.

Safety, Allergenic Potential

Beta glucan materials which are derived from yeast, do not contain any allergenic risk as they contain extremely low protein levels. Beta glucan is a polysaccharide extract of the cell wall, not whole yeast extract. It is not contraindicated for individuals with candidiasis. Generally, refined glucan products are considered hypoallergenic. Now, if you breathe the powder in directly, lungs, nasal membranes, etc. will become irritated as with any powder due to the immune stimulation from the glucan working on the tissues, just as dust, hay or grass will cause irritation also.

"Glucan (Beta-1.3) has been shown to enhance macrophage function dramatically, and to increase nonspecific host resistance to a variety of bacterial, viral, fungal and parasitic infections."
Dept. of Experimental Hematology and Radiation Sciences, Armed Forces Radiobiology Institute

"Beta-1,3 glucan, is a potent macrophage stimulant and is beneficial in the therapy of experimental bacterial, viral and fungal diseases."
Tulane University School of Medicine

"A cascade of interactions and reactions initiated by macrophage regulatory factors can be envisioned to occur …"
Harvard Medical School

Bibliography

Adachi Y, Ohno N, Yadmae T; Activation of murine kupffer cells by administration with gel-forming (1-3)-beta-D-glucan from Grifola frondosa. Biol Pharm Bull 1998;21(3)278-83.

Babineau TJ, Hackford A, Kenler A, et al; A phase II multiple center, double-blind, randomized, placebo-controlled study of three dosages of an immunomodulator (PGG-glucan) in high risk surgical patients. Arch Surg 1994;129(11):1204-10.

Behall KM, Scholfield DJ, Hallfrisch J; Effect of beta glucan level in oat fiber extracts on blood lipids in men and women. J Am Coll Nutr 1997;16(1)46-51.

Bobek P; Galbavy S; The oyster mushroom (Pleurotus ostreatus) effectively prevents the development of atherosclerosis in rabbits] Vyskumny ustav vyzivy, Bratislava. Ceska Slov Farm 1999 Sep;48(5):226-30

Bobek P; Galbavy S; Hypocholesterolemic and antiatherogenic effect of oyster mushroom (Pleurotus ostreatus) in rabbits. Research Institute of Nutrition, Bratislava, Slovak Republic.Nahrung 1999 Oct;43(5):339-42

Bobek P; Ozdin O; Mikus M; Dietary oyster mushroom (Pleurotus ostreatus) accelerates plasma cholesterol turnover in hypercholesterolaemic rat. Physiol Res 1995;44(5):287-91.

Bormann, C., Arrowhead HealthWorks, personal conversation, Dec 20, 2000

Braaten JT; Scott FW; Wood PJ; et al; High beta-glucan oat bran and oat gum reduce postprandial blood glucose and insulin in subjects with and without type 2 diabetes. Diabet Med 1994 Apr;11(3):312-8.

Braaten JT; Wood PJ; Scott FW; Wolynetz MS; etal; Oat beta-glucan reduces blood cholesterol concentration in hypercholesterolemic subjects. Eur J Clin Nutr 1994 Jul;48(7):465-74

Chang, R; Functional properties of edible mushrooms, Nutr Rev (1996), 54(11 Pt 2):S91-93.

Chihara G; Recent progress in immunopharmacology and therapeutic effects of polysaccharides. Dev Biol Stand 1992;77:191-7.

Ebina T, Fujimiya Y; Antitumor effect of a peptide-glucan preparation extracted from Agaricus blazei in a double-grafted tumor system in mice. Biotherapy 1998;11(4):259-65.

Estrada A; Yun CH; Van Kessel A; Li B; Hauta S; Laarveld B; Immunomodulatory activities of oat beta-glucan in vitro and in vivo. Microbiol Immunol 1997;41(12):991-8

Di Renzo L, Yefenof E, Klien E; The function of human NK cells is enhanced by beta glucan, a ligand of CR3 (CD11b/CS18). Eur J Immunol 1991;21(7)1755-8.

Douwes J, Zuidhof A, Doekes G, van Der ZEE S, Wouters I, Marike; Boezen H, Brunekreef B; (1—> 3)-beta-D-glucan and endotoxin in house dust and peak flow variability in children. Am J Respir Crit Care Med. 2000 Oct;162(4 Pt 1):1348-54.

Fujimiya Y, Suzuki Y, Oshiman K, Kobori H, Moriguchi K, et al; Selective tumoricidal effect of soluble proteoglucan extracted from the basidiomycete, Agaricus blazei Murill, mediated via natural killer cell activation and apoptosis. Cancer Immunol Immunother 1998 May;46(3):147-59.

Fujimiya Y,Suzuki Y, Katakura R, Ebina T; Tumor-specific cytocidal and immunopotentiating effects of relatively low molecular weight products derived from the basidiomycete, Agaricus blazei Murill. Anticancer Res 1999 Jan-Feb;19(1A):113-8.

Gardner S, Freyschmidt-Paul P, Hoffmann R, et al; Normalisation of hair follicle morphology in C3H/HeJ alopecia areata mice after treatment with squaric acid dibutylester. Eur J Dermatol 2000 Aug;10(6):443-50.

Gupta MA, Gupta AK, Watteel GN; Stress and alopecia areata: a psychodermatologic study.Acta Derm Venereol 1997 Jul;77(4):296-8.

Hamuro J, Rollinghoff M, Wagner H; beta (1 leads to 3) Glucan-mediated augmentation of alloreactive murine cytotoxic T lymphocytes in vivo. Cancer Res 1978;38(9):3080-5.

Higaki M; Eguchi F; Watanabe Y; A stable culturing method and pharmacological effects of the Agaricus blazei] Lab. of Forest Products Chemistry, Tokyo University of Agriculture, Japan. Nippon Yakurigaku Zasshi 1997 Oct;110 Suppl 1:98P-103P.

Hofer M, Pospisil M. Glucan as a stimulator of hematopoeisis in normal and gamma-irradiated mice. Int J Immunopharmacol 1997;19(19)607-9.

Hoffman OA, Olson EJ, Limper AH Fungal beta-glucans modulate macrophage release of tumor necrosis factor-alpha in response to bacterial lipopolysaccharide. Immunol Lett. 1993 Jul;37(1):19-25.

Ito H, Shimura K, Itoh H, Kawade M; Antitumor effects of a new polysaccharide-protein complex (ATOM) prepared from Agaricus blazei (Iwade strain 101) "Himematsutake" and its mechanisms in tumor-bearing mice. Anticancer Res 1997 Jan-Feb;17(1A):277-84.

Itoh H, Ito H, Amano H, Noda H; Inhibitory action of a beta-D-glucan-protein complex (F III-2-b) isolated from Agaricus blazei Murill ("himematsutake") on Meth A fibrosarcoma-bearing mice and its antitumor mechanism. Mie University, Japan. Jpn J Pharmacol 1994 Oct;66(2):265-71.

Kawagishi H, Inagaki R, Kanao T, Mizuno T, et al; Fractionation and antitumor activity of the water-insoluble residue of Agaricus blazei fruiting bodies. Carbohydr Res 1989 Mar 15;186(2):267-73

Kim HM, Han SB, Oh GT, Kim YH, et al; Stimulation of humoral and cell mediated immunity by polysaccharide from mushroom Phellinus linteus. Int J Immunopharmacol 1996 May;18(5):295-303.

Kulicke WM, Lettau AI, Thielking H; Correlation between immunological activity, molecular mass and molecular structure of different (1-3)-beta-D-glucan. Carbohydr Res 1997;297(2)135-42.

Lia A; Hallmans G; Sandberg AS; Sundberg B; et al; Oat beta-glucan increases bile acid excretion and a fiber-rich barley fraction increases cholesterol excretion in ileostomy subjects.Am J Clin Nutr 1995 Dec;62(6):1245-51

Liang J, Melican D, Cafro L et al; Enhanced clearance of a multiple antibiotic resistant Staphylococcus aureus in rats treated with PBB-glucan is associated with increased leukocyte counts and increased neutrophil oxidative burst activity. Int J Immunopharmacol 1998;20(11)595-614.

Lovegrove JA; Clohessy A; Milon H; Williams CM; Modest doses of beta-glucan do not reduce concentrations of potentially atherogenic lipoproteins. Am J Clin Nutr 2000 Jul;72(1):49-55

Mizuno M, Minato K, Ito H, et al; Anti-tumor polysaccharide from the mycelium of liquid-cultured Agaricus blazei mill.Biochem Mol Biol Int 1999 Apr; 47(4):707-14.

Mizuno M, Morimoto M, Minato K, Tsuchida H; Polysaccharides from Agaricus blazei stimulate lymphocyte T-cell subsets in mice. Biosci Biotechnol Biochem 1998 Mar;62(3):434-7.

Ohno N, Asada N, Adachi Y et al; Enhancement of LPS triggered TNF-alpha (tumor necrosis factor-alpha) production by (1-3) beta-D-glucan in mice. Biol Pharm Bull 1995;8(1):126-33.

Osaki Y, Kato T, Yamamoto K, Okubo J,; Antimutagenic and bactericidal substances in the fruit body of a Basidiomycete Agaricus blazei, Jun-17] Tokyo College of Pharmacy, Japan. Yakugaku Zasshi 1994 May; 114(5): 42-50.

Patchen ML, Liang J, Vaudrain T, et al; Mobilization of peripheral progenitor cells by Betafectin PGG-glucan alone and in combination with granulocyte colony-stimulating factor. Stem Cells 1998;16(3)208-17.

Paulik S, Svrcek S, Mojzisova J, Durove A, Benisek Z, Huska MThe immunomodulatory effect of the soluble fungal glucan (Pleurotus ostreatus) on delayed hypersensitivity and phagocytic ability of blood leucocytes in mice.Zentralbl Veterinarmed [B]. 1996 May;43(3):129-35.

Paulik S, Levkutova M, Bajova V, Benko G, Harvan M[The effect of primary BHV-1 infection on the dynamics of T and B lymphocytes in the peripheral blood and levels of specific serum antibodies in calves treated with glucan]. Vet Med (Praha). 1993;38(8):477-83.

Paulik S, Svrcek S, Huska M, Mojzisova J, Durove A, Benisek Z [The effect of fungal and yeast glucan and levamisole on the level of the cellular immune response in vivo and leukocyte phagocytic activity in mice].Vet Med (Praha). 1992 Dec;37(12):675-85.

Paulik S, Bajova V, Mojzisova J [The effect of administration of non-soluble fungal glucan on the level of delayed skin hypersensitivity and phagocytic activity of leukocytes in the blood of calves].Vet Med (Praha). 1992 Nov;37(11):577-85.

Penna C, Dean PA, Nelson H; Pulmonary metastases neutralization and tumor rejection by in vivo administration of beta glucan and bispecific antibody. Int J Cancer 1996; 65(3)377-82.

Ross GD; Vetvicka V; Yan J; Xia Y; Vetvickova J Therapeutic intervention with complement and beta-glucan in cancer. Department of Pathology, University of Louisville, KY Immunopharmacology 1999 May;42(1-3):61-74

Rylander R, Lin R(1—>3)-beta-D-glucan - relationship to indoor air-related symptoms, allergy and asthma. Toxicology. 2000 Nov 2;152(1-3):47-52.

Sakurai T; Ohno N; Yadomae T; Effects of fungal beta-glucan and interferon-gamma on the secretory functions of murine alveolar macrophages.   Laboratory of Immunopharmacology of Microbial Products, School of Pharmacy, Tokyo University of Pharmacy and Life Science, Japan. J Leukoc Biol 1996 Jul;60(1):118-24.

Sakurai T, Hashimoto K, Suzuki I, et al; Enhancement of murine alveolar macrophage functions by orally administered beta glucan. Int J Immunopharmacol 1992;14(5)821-30.

See, D, Agaricus blazei Murill Testing, Institute of Longevity Medicine, Huntington Beach, CA . 1999: Oct 27.

Soltys J; Quinn MT; Modulation of endotoxin- and enterotoxin-induced cytokine release by in vivo treatment with beta-(1,6)-branched beta-(1,3)-glucan. Department of Veterinary Molecular Biology, Montana State University, Bozeman Infect Immun 1999 Jan;67(1):244-52.

Tamuro J, Rollinghoff M, Wagner H; beta (1 leads to 3) Glucan-mediated augmentation of alloreactive murine cytotoxic T lymphocytes in vivo. Cancer Res 1978;38(9):3080-5.

Tappy L; Gugolz E; Wursch P; Effects of breakfast cereals containing various amounts of beta-glucan fibers on plasma glucose and insulin responses in NIDDM subjects. Diabetes Care 1996 Aug;19(8):831-4.

Thornton BP, Vetvicka V, Pitman M, et al; Analysis of the sugar specificity and molecular location of the beta glucan binding lectin site of complement receptor type 3. J Immunol 1996;156(3):1235-46.

Tsujinaka T, Yokota M, Kambayashi J, et al; Modification of septic processes by beta glucan administration. Eur Surg Res 1990;22(6):340-6.

Tzianabos AO, Gibson FC 3rd, Cisneros RL, Kasper DL.; Protection against experimental intra-abdominal sepsis by two polysaccharide immunomodulators. J Infect Dis 1998;178(1)200-6.

Uusitupa MI; Miettinen TA; Sarkkinen ES; et al; Lathosterol and other non-cholesterol sterols during treatment of hypercholesterolaemia with beta-glucan-rich oat bran. Eur J Clin Nutr 1997 Sep;51(9):607-11.

Wood PJ; Beer MU; Butler G; Evaluation of role of concentration and molecular weight of oat beta-glucan in determining effect of viscosity on plasma glucose and insulin following an oral glucose load. Br J Nutr 2000 Jul;84(1):19-23.

Wursch P; Pi-Sunyer FX; The role of viscous soluble fiber in the metabolic control of diabetes. A review with special emphasis on cereals rich in beta-glucan. Nestle Research Centre, Lausanne, Switzerland. Diabetes Care 1997 Nov;20(11):1774-80.
Zelenkova, Dermatology department of NsP Svidnik, Slovak republic.

INDEX

ABOUT THE AUTHOR

Beth M. Ley, Ph.D., has been a science writer specializing in health and nutrition for over 12 years and has written over a dozen health related books, including the best sellers, ***DHEA: Unlocking the Secrets to the Fountain of Youth*** and ***MSM: On Our Way Back to Health With Sulfur***. She wrote her own undergraduate degree program and graduated in Scientific and Technical Writing from North Dakota State University in 1987 (combination of Zoology and Journalism). Beth has her masters (1997) and doctoral degrees (1999) in Nutrition.

Beth lives in the Minnesota lakes country. She is dedicated to God and to spreading the health message. She enjoys spending time with her Dalmatian, exercises on a regular basis, eats a vegetarian, low-fat diet and takes anti-aging supplements.

Memberships: American Academy of Anti-aging, New York Academy of Sciences, Oxygen Society.

ORDER THESE GREAT BOOKS
FROM BL PUBLICATIONS!

Immune System Control
Colostrum & Lactoferrin
Beth M. Ley, Ph.D. 200 pages, $12.95 ISBN 1-890766-11-9
Get the indepth and detailed FACTS about colostum and lactoferrin! Testimonials and much more! Also features a special product selection guide! Fully referenced/Indexed

Fading: One family's journey with a woman silenced by Alzheimer's - also with Preventative, Nutritional and Psychological Help for the Families and Patients with Alzheimer's Disease

Frances Kraft, Dr. Barry Kraft, Dr. Beth Ley, 2000, 200 pgs. $12.95

An excellent inspirational and resource book for all families and caretakers.

Aspirin Alternatives:
The Top Natural Pain-Relieving Analgesics
Raymond Lombardi, D.C., N.D., C.C.N., 1999, 160 pages, $8.95

This book discusses analgesics and natural approaches to pain. Ibuprofen and acetaminophen are used for pain-relief, but like all drugs, there is a risk of side effects and interactions, Natural alternatives are equally effective and in many cases preferable because they may help treat the underlining problem rather than simply treating a symptom.

Vinpocetine: Boost Your Brain w/ Periwinkle
Extract! Beth M. Ley, Ph.D. 2000, 48 pgs. $4.95

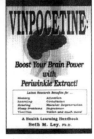

This herbal extract benefits: Memory, attention and concentration, learning, circulation, hearing, insomnia, depression, tinnitus, vision & more! Vinpocetine increases circulation in the brain and increases metabolism in the brain.

Coenzyme Q10: All Around
Nutrient for All-Around Health!
Beth M. Ley-Jacobs, Ph.D., 1999, 60 pages, $4.95

CoQ10 is found in every living cell. With age, insufficient levels become more common, putting us at serious risk of illness and disease. Protect and strengthen the cardiovascular system; benefit blood pressure, immunity, fatigue, weight problems, Alzheimer's, Parkinson's, Huntington's, gum-disease and slow aging.

MSM: On Our Way Back To Health With
Sulfur Beth M. Ley, 1998, 40 pages, $3.95

MSM (methyl sulfonyl methane), is a rich source of organic sulfur, important for connective tissue regeneration. Beneficial for arthritis and other joint problems, allergies, asthma, skin problems, TMJ, periodontal conditions, pain relief, and much more! Includes important "How to use" directions.

How to Fight Osteoporosis & Win: The Miracle of MCHC
Beth M. Ley, 80 pgs. $6.95

Find out if you are at risk for osteroporosis and what to do to prevent and reverse it. Get the truth about bone loss, calcium, supplements, foods, MCHC & much more! Find out what supplements can help you most!

Nature's Road to Recovery: Nutritional Supplements for The Social Drinker, Alcoholic & Chemical-Dependent
Beth M. Ley-Jacobs, Ph.D., 1999, 72 pages, $5.95

Recovery is much more than abstinence. Cravings, depression, memory loss, liver problems, vascular problems, sexual problems, sleep problems, nutritional deficiencies and common health problems which can benefit from 5-HTP, DHA, phospholipids, St. John's Wort, antioxidants, etc.

DHA: The Magnificent Marine Oil
Beth M. Ley-Jacobs, Ph.D., 1999, 120 pages, $6.95

Individuals commonly lack this essential Omega-3 fatty acid so important to the brain, vision, and immune system and much more. Memory, depression, ADD, addiction disorders, inflammatory disorders, skin problems, schizophrenia, elevated blood lipids, etc., benefit from DHA.

Marvelous Memory Boosters Beth M. Ley, Ph.D. 2000,
32 pages, $3.95

Certain nutrients & phytochemicals (Alpha GPC, Vinpocetine, Huperzine-A, Pregnenolone, Phospholipids, DHA, Bacopa Monniera, Ginkgo Biloba, etc.) improve short & long term memory, increase mental acuity & concentration, improve learning abilities & mental stamina, reduce fatigue, improve sleep, mood, vision & hearing.

Bilberry & Lutein: The Vision Enhancers
Beth M. Ley, Ph.D. 40 pgs. $4.95

Find out how bilberry and lutein, important antioxidants specifically for the eyes, can help improve your vision and ward off common eye problems including macular degeneration, cataracts, glaucoma, retinopathy, etc.

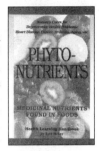

Phyto-Nutrients: Medicinal Nutrients Found in Foods, Beth M. Ley, 40 pgs. $3.95

Learn about special components in our foods which protect our health by fighting off disease & aging! Learn about onions, garlic, flax, bilberry, green tea, red wine, rosemary, cruciferous vegetables, cayenne pepper, ginger, soybeans, avacados, beets, cranberries, sweet potatoes, amaranth, etc.

Colostrum: Nature's Gift to The Immune System,
Beth M. Ley, 80 pages, $5.95

An earlier edition - Colostrum, "first milk," is rich in immuno-factors such as antibodies, growth factors, lactoferrin, etc., which can boost & support the immune system of everyone!

DHEA: Unlocking the Secrets to the Fountain of Youth - 2nd Ed. Richard Ash, M.D., & Beth M. Ley, 320 pgs. $14.95

Excellent resource guide for practicioners and patients. Find how how to use the famed "anti-aging hormone," DHEA, safely, without side effects, to reverse aging & treat & prevent disease.

Natural Healing Handbook
Beth M. Ley, 320 pgs. $14.95

Excellent, easy -to-use reference book with natural health care remedies for all your healthcare concerns. A book you will use over & over again!

TO PLACE AN ORDER:

___ *Aspirin Alternatives: The Top Natural Pain-Relieving Analgesics* (Lombardi) $8.95

___ *Discover the Beta Glucan Secret! (Ley)* $3.95

___ *Bilberry & Lutein: The Vision Enhancers! (Ley)* $4.95

___ *Calcium: Fossilized Coral* (Ley) $4.95

___ *Castor Oil: Its Healing Properties* (Ley) $3.95

___ *Dr. John Willard on Catalyst Altered Water* (Ley) $3.95

___ *Co Q10: All-Around Nutrient for All-Around Health* (Ley) .$4.95

___ *Colostrum: Nature's Gift to the Immune System* (Ley) $5.95

___ *DHA: The Magnificent Marine Oil* (Ley Jacobs) $6.95

___ *DHEA: Unlocking the Secrets of the Fountain of Youth- 2nd Edition* (Ash-Ley) $14.95

___ *Fading: One family's journey with a women silenced by Alzheimer's* (Kraft) $12.95

___ *God Wants You Well* (Ley) $14.95

___ *Health Benefits of Probiotics* (Dash) $4.95

___ *How to Fight Osteoporosis and Win!* (Ley) $6.95

___ *Immune System Control-Colostrum & Lactoferrin* (Ley) $12.95

___ *Marvelous Memory Boosters* (Ley) $3.95

___ *Medicinal Mushrooms:* **Agaricus blazei Murill** *(Ley)* .$4.95

___ *MSM: On Our Way Back to Health W/ Sulfur* (Ley) $3.95

___ *Natural Healing Handbook* (Ley) $14.95

___ *Nature's Road to Recovery: Nutritional Supplements for the Alcoholic & Chemical Dependent* (Ley Jacobs) $5.95

___ *PhytoNutrients: Medicinal Nutrients Found in Foods* (Ley) $3.95

___ *The Potato Antioxidant: Alpha Lipoic Acid* (Ley) $6.95

___ *Vinpocetine: Revitalize Your Brain w/ Periwinkle Extract!* (Ley) $4.95

___ *Yesterday, Today & Tomorrow: Take the High Road to Health with Rod Burreson* (Burreson) $24.95

Book subtotal $ _____   Please add $3.00 for shipping

TOTAL $_____

Send check or money order to:

BL Publications 14325 Barnes Drive Detroit Lakes, MN 56501

Credit card orders call toll free: 1-877-BOOKS11
visit: www.blpbooks.com